IIW Collection

Series editor

IIW International Institute of Welding,
ZI Paris Nord II, Villepinte, France

About the Series

The IIW Collection of Books is authored by experts from the 59 countries participating in the work of the 23 Technical Working Units of the International Institute of Welding, recognized as the largest worldwide network for welding and allied joining technologies.

The IIW's Mission is to operate as the global body for the science and application of joining technology, providing a forum for networking and knowledge exchange among scientists, researchers and industry.

Published books, Best Practices, Recommendations or Guidelines are the outcome of collaborative work and technical discussions-they are truly validated by the IIW groups of experts in joining, cutting and surface treatment of metallic and non-metallic materials by such processes as welding, brazing, soldering, thermal cutting, thermal spraying, adhesive bonding and microjoining. IIW work also embraces allied fields including quality assurance, non-destructive testing, standardization, inspection, health and safety, education, training, qualification, design and fabrication.

More information about this series at http://www.springer.com/series/13906

Gary B. Marquis · Zuheir Barsoum

IIW Recommendations for the HFMI Treatment

For Improving the Fatigue Strength of Welded Joints

INTERNATIONAL INSTITUTE OF WELDING
A world of joining experience

Springer

Gary B. Marquis
Aalto University
Helsinki
Finland

Zuheir Barsoum
KTH Royal Institute of Technology
Stockholm
Sweden

ISSN 2365-435X ISSN 2365-4368 (electronic)
IIW Collection
ISBN 978-981-10-9633-4 ISBN 978-981-10-2504-4 (eBook)
DOI 10.1007/978-981-10-2504-4

Printed on acid-free paper

This Springer imprint is published by Springer Nature
The registered company is Springer Nature Singapore Pte Ltd.
The registered company address is: 152 Beach Road, #22-06/08 Gateway East, Singapore 189721, Singapore

Contents

About the Authors

Gary B. Marquis is a professor of applied mechanics at the Aalto University, Helsinki, Finland, where he currently serves as a dean for school of engineering. Within IIW, he has served as a chairman for Commission XIII (*Fatigue of welded components and structures*) and is currently the president of IIW.

Zuheir Barsoum is an associate professor of lightweight structures at KTH Royal Institute of Technology, Stockholm, Sweden. Within IIW, he is currently leading the XIII-WG2 on *Techniques for improving the fatigue strength of welded joints, components and structures* and serving as vice chair for Commission XIII.

Nomenclature

D	Damage sum for variable amplitude loading
f_y	Yield strength
FAT	IIW fatigue class, i.e., the nominal stress range in MPa corresponding to 95 % survival probability at 2×10^6 cycles to failure (a discrete variable with 10–15 % increase in stress between steps)
$f(t)$	IIW thickness correction factor
k_R	Strength reduction factor for stress ratio, $1 \geq k_R > 0$
K	Stress concentration
m	Slope of the S-N line $1 \times 10^4 \leq N < 1 \times 10^7$ cycles
m'	Slope of the S-N line 1×10^7 cycles $\leq N$
L	Characteristic length used to compute $f(t)$
N	Fatigue cycles
R	Stress ratio
t	Plate thickness
X_N	Improvement factor in life for HFMI-treated welds at $\Delta\sigma$ equal to the FAT class of the as-welded joint: $N_f = X_N \times 2 \times 10^6$
ρ	Weld toe radius
σ	Stress
$\Delta\sigma$	Stress range

Subscripts

eff	Effective
f	Failure (cycles) or fictitious (weld toe radius)
k	Corresponding to the knee point of the S-N curve
S	Structural hot spot stress

max	Maximum value: during one cycle for constant amplitude loading or during one repetition of the spectrum for variable amplitude loading
min	Minimum allowable
nom	Nominal
w	Value computed using the effective notch method

Abstract

High-frequency mechanical impact (HFMI) has emerged as a reliable, effective, and user-friendly method for post-weld fatigue strength improvement technique for welded structures. This guideline presents recommendations on proper treatment procedures, quality control measures, and fatigue strength improvement assessment based on nominal, hot spot, and effective notch stress methods. Recommendations on effect of loading conditions, variable amplitude loading, low cycle fatigue, and consideration of low-stress concentration details are also given. The guideline is applicable to steel structures of plate thicknesses of 5–50 mm and for yield strength ranging from 235 to 960 MPa.

Keywords Peening · Weld toe · Fatigue improvement · High-strength steels · Fatigue strength · Stress analysis

IIW Recommendations on High Frequency Mechanical Impact (HFMI) Treatment for Improving the Fatigue Strength of Welded Joints

Abstract High-frequency mechanical impact (HFMI) has emerged as a reliable, effective and user-friendly method for post-weld fatigue strength improvement technique for welded structures. This guideline presents recommendations on proper treatment procedures, quality control measures and fatigue strength improvement assessment based on nominal, hot spot and effective notch stress methods. Recommendations on effect of loading conditions, variable amplitude loading, low cycle fatigue and consideration of low stress concentration details are also given. The guideline is applicable to steel structures of plate thicknesses of 5–50 mm and for yield strength ranging from 235 to 960 MPa.

Keywords Peening · Weld toe · Fatigue improvement · High-strength steels · Fatigue strength · Stress analysis

1 Introduction

Weld toe improvement methods have been widely investigated and have in most cases been found to give substantial increases in fatigue strength. However, there are large variations in the actual improvements achieved, and the results obtained by various methods are not always ranked in a consistent manner. One explanation for the observed variations is the lack of standardization of the optimum method of application, but variations in the material, type of loading and type of test specimens may also have influenced the results. The effectiveness of the treatment also depends heavily on the skill of the operator.

High Frequency Mechanical Impact (HFMI) has emerged as a reliable, effective and user-friendly method for post-weld fatigue strength improvement technique for welded structures. This recommendation presents an overview of HFMI techniques existing today in the market and their proper procedures, quality assurance measures and documentation [1]. Due to differences in HFMI tools and the wide variety of potential applications, certain details of proper treatments and quantitative

G.B. Marquis and Z. Barsoum, *IIW Recommendations for the HFMI Treatment*,
IIW Collection, DOI 10.1007/978-981-10-2504-4_1

quality control measures are presented generally. An example of procedure specification as a quality assurance measure is given in Appendix 1.

Moreover, the recommendation presents procedures for the fatigue life assessment of HFMI-improved welded joints based on nominal stress, structural hot spot stress and effective notch stress [2]. The recommendation also considers the observed extra benefit that has been experimentally observed for HFMI-treated high-strength steels. Proposals on the effect of the effect of loading conditions like high mean stress fatigue cycles, variable amplitude loading and large amplitude/low cycle fatigue cycles are given. Special considerations for low stress concentration welded joints are also given. In order to demonstrate the use of the guideline, several fatigue assessment examples are provided in Appendix 2.

1.1 Background

In 2007 Commission XIII: Fatigue of Welded Components and Structures approved the best practice recommendations concerning post-weld treatment methods for steel and aluminum structures [3]. This recommendation covers four commonly applied post weld treatment methods, burr-grinding, TIG re-melting (i.e. TIG dressing), hammer peening and needle peening.

Simultaneous with the development of the 2007 recommendations, an increasing number of presentations within Commission XIII focused on high frequency mechanical impact (HFMI) as a means of improving the fatigue strength of welded structures. During the period 2002–2012, 46 IIW Commission XIII documents reporting HFMI technology or experimental studies involving HFMI-based fatigue strength improvement were presented. HFMI has emerged as a reliable, effective and user-friendly method for post-weld fatigue strength improvement technique for welded structures. A great deal of experimental evidence has demonstrated that HFMI can significantly improve the fatigue strength of welded structures.

This guideline attempts to summarize the knowledge gained both within the IIW and presented in the open international literature. It covers procedure and quality assurance related items such as relevant equipment, proper application procedures, material requirements, safety, training requirements for operators and inspectors, quality control measures, documentation and fatigue design proposal for HFMI treatment of welded steel structures. All HFMI devices have unique features and type of structure being treated will greatly influence the optimal treatment procedures. For this reason, the current guidelines are intended to provide only general recommendations especially with respect to operator training, procedures and inspection. Specific operator training is provided by the tool manufacturers. Specific treatment and requirements can normally be developed in cooperation with the HFMI device manufacturer. The recommendations in this guideline are dedicated to structural steels only. Studies on HFMI treated welds in aluminum and stainless steel structures exist to a limited extend where fatigue life improvements are observed. Moreover, additional data are necessary to establish recommendations for design.

1.2 Purpose

The current IIW recommendations on post weld fatigue life improvement of steel and aluminum welded structures [3] covers several important life extension techniques. The recommendations by Haagensen and Maddox [3] cover the nominal and structural hot spot stress assessment methods for improved welded joints for limited steel grades. The fast development of new and higher strength steels and HFMI techniques motivates the development of new guidelines.

It is hoped that these international guidelines are relevant to a variety of HFMI technologies which are applicable to many industrial sectors facing needs of improving the fatigue strength of welded structures. The guidelines can also be used as means of verifying the effectiveness of new HFMI equipment as it comes to the market. The HFMI treatment of welded joints described in these recommendations aims primarily to; increase the fatigue strength of new structures, repair or upgrading of existing structures. The recommendation also offers the possibility of upgrading of new/existing of structures by utilizing high-strength steels.

1.3 Overview of Post Weld Improvement Techniques

An important step in order to increase the fatigue life of welded components is to apply good design practice which can be realizes in several ways e.g. use joints with low stress concentration factor (butt welds instead of fillet welds), and/or place welds in areas of low stresses. Increased fatigue life is also achieved by high quality fabrication [4–7] e.g. by proper selection of material, welding process, weld preparation and groove geometry. New weld quality system for high quality production exist and are used successfully [6]. In case of none of above mentioned steps fulfill the fatigue life target weld improvement techniques can be used during or post production.

Post weld improvement techniques could generally be divided in two main groups depending on how the improvement is achieved; by local weld geometry modification e.g. local stress peaks are reduced and surface quality is improved. The second main group is the residual stress techniques where the improvement is to reduce the tensile residual stress in the welds and in many cases introduce compressive residual stresses by e.g. work hardening or phase changes. Figure 1 gives an overview of different improvement techniques on the market today.

Burr-grinding and TIG re-melting are generally classified as geometry improvement techniques for which the primary aim is eliminate weld toe flaws and to reduce the local stress concentration by achieving a smooth transition between the plate and the weld face. Hammer peening and needle peening are classified as residual stress modification techniques which eliminate the high tensile residual stress in the weld toe region and induce compressive residual stresses at the weld toe. These methods also result in a reduced stress concentration at the weld toe.

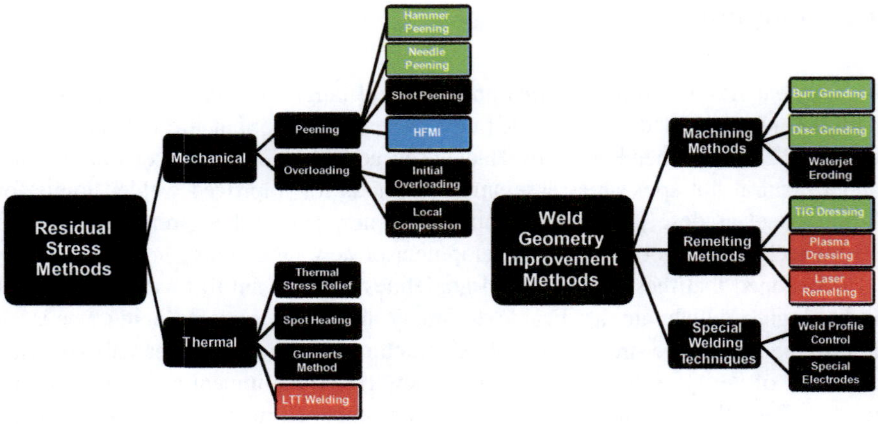

Fig. 1 Overview of different weld improvement techniques. *Green* is covered by IIW recommendations [3], red is planned/in progress and *blue* is the current guideline

Table 1 Example of weld improvement methods and their main effects

Method	Weld geometry improvement		Mechanical effects
	Increasing and smoothing transition	Eliminates defects	Induces compressive residual stresses
Grinding	x	x	–
TIG-remelting	x	x	–
Shot peening (blasting)	–	–	x
Hammer/needle peening	x	x	x
HFMI	x	x	x

HFMI techniques improves the local weld geometry and surface quality and at the same time introduces high compressive residual stresses [8], Table 1 gives an overview of the major beneficial improvement different techniques can achieve.

In the recommendations by Haagensen and Maddox [3], recommendations are given for four different improvement techniques; Burr grinding, TIG dressing, Hammer-and Needle peening. The later aforementioned are considered resulting in similar fatigue life improvement as the collection of HFMI techniques, e.g. the fatigue life improvements claimed for HFMI improved welds, in Sect. 5 in this document is also applicable for Hammer-and Needle peening. However, for the procedure, equipment and quality control of these two techniques one should consult the recommendations by Haagensen and Maddox [3].

2 High Frequency Mechanical Impact (HFMI)

In the past decade, there has been a steady increase in the number of HFMI peening equipment manufacturers and service providers. In 2010 Commission XIII of the IIW introduced the term high frequency mechanical impact (HFMI) as a generic term to describe several related technologies. Alternate power sources are employed, for example, ultrasonic piezoelectric elements, ultrasonic magnetostrictive elements or compressed air. In all cases, however, the working principal is identical: cylindrical indenters are accelerated against a component or structure with high frequency (about 90 Hz). The beneficial effect is mainly because of the impacted energy per indentation; which is determined by velocity (e.g. frequency) and driven mass. A single indent should have an average force in the kN-range. The impacted material is highly plastically deformed causing changes in the material microstructure and the local geometry as well as the residual stress state in the region of impact.

Various names have been used in literature to describe the devices: ultrasonic impact treatment (UIT) [9], ultrasonic peening (UP) [10], ultrasonic peening treatment (UPT) [11, 12], high frequency impact treatment (HiFiT) [13], pneumatic impact treatment (PIT) [14] and ultrasonic needle peening (UNP) [15, 16]. Figure 2 shows typical weld profiles in the as-welded condition and following HFMI treatment. In comparison to hammer peening, the operation is considered to be more user-friendly and the spacing between alternate impacts on the work piece is very small resulting in a finer surface finish. While HFMI can be considered as environmental friendly, safe and relatively easy to apply, operators must still exercise safe work practice and understand the equipment and the nature of the post-weld operation which is being imparted to a welded structure.

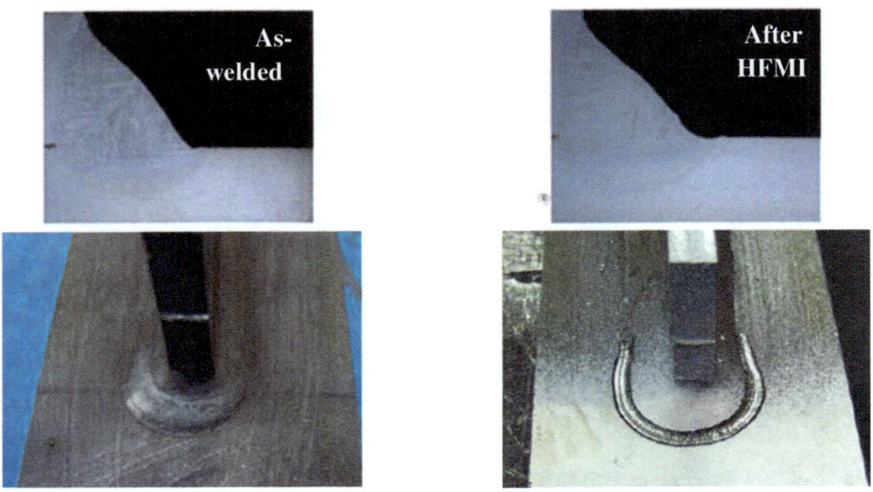

Fig. 2 Typical weld toe profile in the as-welded condition and following HFMI treatment

2.1 HFMI Equipment

Numerous new HFMI devices have been developed during the past 10 years and the number continues to increase. Figure 3 shows some of the HFMI devices that are in use worldwide today. A recent round robin exercise [17] have identified several HFMI tools which, when properly used, provide the same degree of fatigue life improvement. Also hammer-and needle peening have showed to result in similar degrees of fatigue life improvement [18]. Similar devices can be assigned to this group if they have the same operating principal and are objectively tested and are shown to have the same reliable and beneficial effect on the fatigue strength of welds as in the proposed guideline.

Ultrasonic devices consist of a power unit and tool. These normally require compressed air or circulating water to control the temperature of the tool. Other devices known to the authors are pneumatic. The indenters are high strength steel

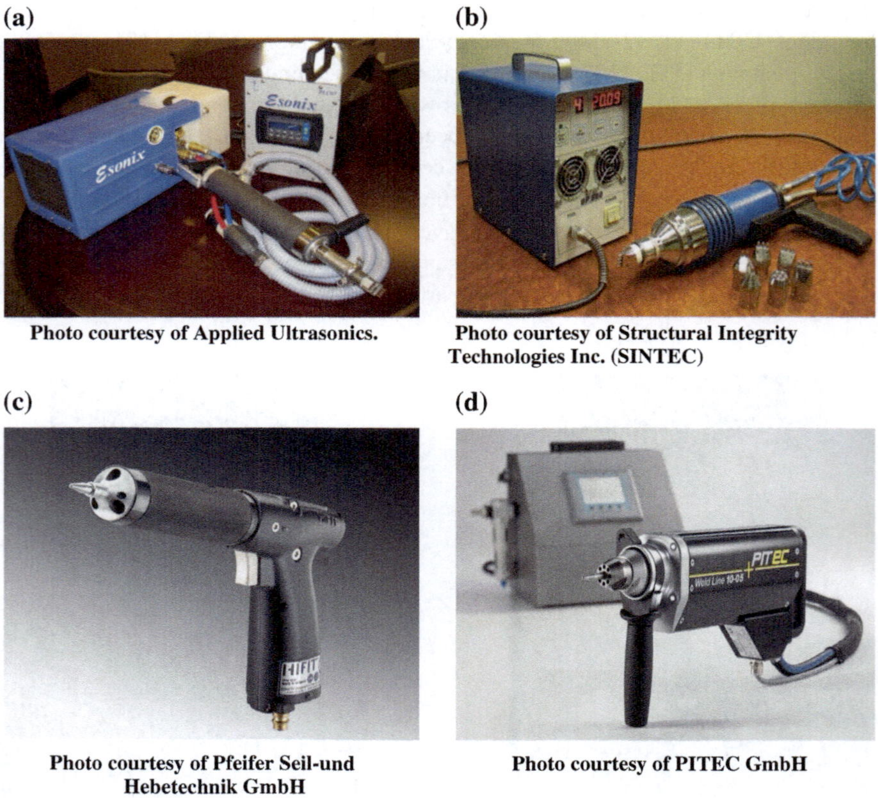

Fig. 3 Examples of HFMI devices available worldwide. **a** Ultrasonic impact treatment (UIT), **b** ultrasonic peening (UP), **c** high frequency impact treatment (HiFiT) and **d** pneumatic impact treatment (PIT)

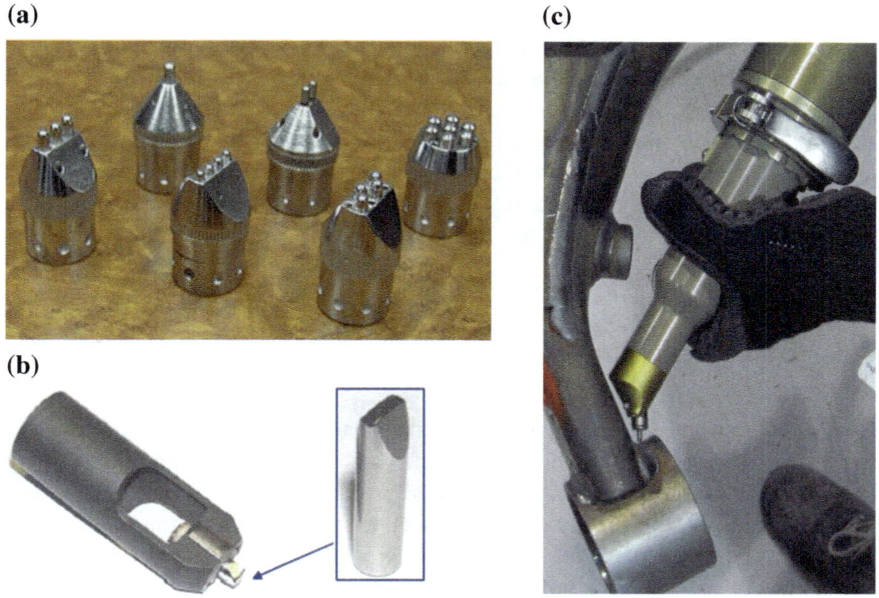

Photo courtesy of Structural Integrity Technologies Inc. (SINTEC)

Photo courtesy of SONATS Europe Technology Group

Fig. 4 **a** Examples of indenter sizes and configurations, **b** a double radius indenter and **c** angled indenter configuration

(HSS) cylinders and manufacturers have customized the effectiveness of their own tools by using indenters with different diameters, tip geometries or multiple indenter configurations. Indenters are consumable items which from time to time require replacement or refurbishment. Figure 4 shows several examples of indenter sizes and configurations which are available.

3 HFMI Procedures

3.1 Operator Training

When delivering new equipment, tool manufacturers normally provide 1–2 days of operator training. As new applications arise, tool manufactures can provide specialized training or customized procedure specifications. In some cases HFMI treatment of structures with curvilinear weld toes, e.g., weld toes in trusses fabricated from circular hollow sections, has proven to be very demanding and will require more expertise than for treating long straight welds or short weld corners. Because HFMI is normally specified as a fatigue strength improvement technology for new structures or during repair and retrofitting operations, it is always essential to consult fatigue experts to ensure that all critical regions in a structure are identified and properly treated. Most fatigue loaded structures will normally have only a limited number of locations that are critical from a fatigue point of view. Proper identification of these regions is also important to avoid extra costs and treatment of regions which are not fatigue critical. Additionally, the possibility of a failure starting at some other location must always be considered. For instance, if the failure origin is merely shifted from the weld toe to the root there may be no significant improvement in fatigue life.

3.2 Weld Preparation

The weld cap and adjacent parent material shall be fully de-slagged and wire brushed or ground to remove all traces of oxide, scale, spatter and other foreign material. HFMI treatment of a convex weld profile or of a weld with a large weld angle can cause the plastically deformed metal to fold over the original weld toe and leave a crack-like lap feature that resembles a cold lap. The weld bead profile should meet the acceptance limits for weld profile quality level B in ISO 5817 [19]. This requirement does not imply that the weld must fulfill all quality level B criteria in ISO 5817; only weld profile-related quality criteria need to be evaluated. These include Undercuts (imperfection 1.7), Excessive overfill (imperfection 1.19), Excessive concavity (imperfection 1.10) and Overlaps (imperfection 1.13). If the weld profile does not comply with these acceptance limits, light grinding before treatment may be desired. It should be noted, however, that HFMI treatment is most effective when the weld toe region itself is treated. Thus, grinding operations which make it difficult for the HFMI operator to distinguish the exact location of the weld toe should be avoided. Decisions on the need for weld grinding and the proper grinding procedure should be agreed on with an experienced HFMI operator. If the weld toe is treated insufficiently, the increase in fatigue strength may not be feasible.

The need for proper weld profile before HFMI is illustrated in Fig. 5a which illustrates the formation of a crack-like defect due to improper contact between the indenter and weld toe. Surface inspection of such a defect reveals a dark crack-like line in the middle of the otherwise smooth and shiny HFMI groove as seen in Fig. 5b. Figure 5c shows section micrographs of these defects. The same type of flaw has been observed in welds with adequate profiles but with improper indenter selection or too severe treatment, i.e. too many passes over the same region. These imperfections are somewhat unavoidable, but will not necessarily lower the improvement in fatigue strength. For specific applications, it may be needed to consult with the HFMI tool manufacturer in order to select the proper treatment procedures and optimal indenter configuration to avoid crack-like defects.

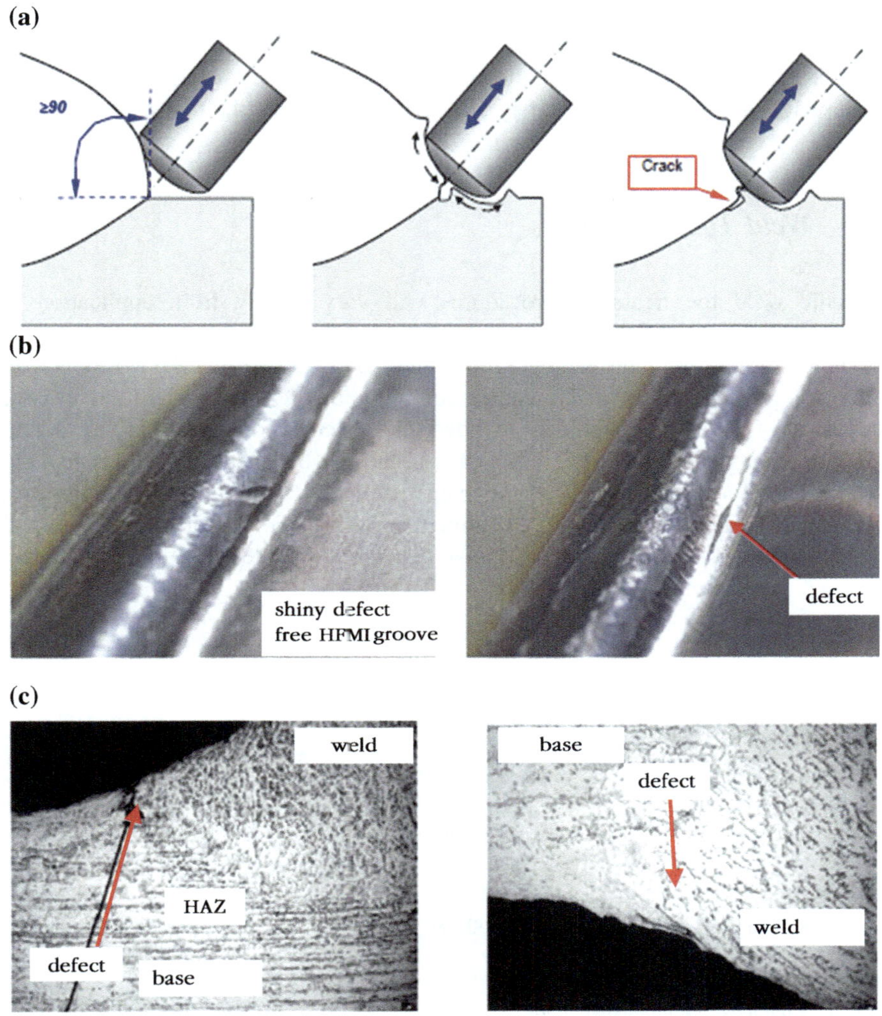

Fig. 5 **a** Potential introduction of crack-like defect due to HFMI treatment of a weld with a steep angle or with too large of an indenter and **b** resulting groove for a properly treated (*left*) and improperly treated weld toe (*right*) and **c** micrographs of the induced crack-like defects due to improper HFMI treatment [18]

3.3 Safety Aspects

Noise and vibration from HFMI is significantly less than for more traditional peening equipment. HFMI treatment can be a noisy operation and it is essential that the operator and others working in the vicinity should use ear protection. Normal protective clothing for working in a fabrication shop is adequate but it should include approved eye protection. Vibration from HFMI equipment is usually low enough so that continuous operation is permitted without restriction during a normal 8 h work shift. If the vibration of the specific HFMI tool has not been determined, it may be needed to limit the amount of time per day for performing HFMI treatment. Equipment-specific safety issues are provided by the tool manufacturers.

3.4 *Weld Toe Treatment*

Specific weld toe treatment procedures will vary greatly from application to application and depending on the tool being used. Thus, only general topics can be covered. Table 2 provides example procedure parameters for two HFMI tools with alternate power sources and indenter configurations see also Fig. 6. Excessive treatment of a weld toe should be avoided, moreover as concluded by Yekta et al. [20], the treatment process is fairly robust in the sense that under- or over-treating within the investigated range of treatment parameter variations still results in a significant fatigue life increase. The American Association of State Highway and Transportation Officials (AASHTO) have developed sample procedures which can be used as examples [21].

Table 2 Sample treatment procedure parameters for two HFMI tools

Parameter	HFMI tool	
	High frequency impact treatment (HiFIT) [22]	Ultrasonic impact treatment (UIT) [21, 23]
Power source	Pneumatic	Ultrasonic magnetostrictive
Number of indenters	1	1–4
Angle of the axis of the indenters with respect to the plate surface, ϕ (see Fig. 6)	60°–80°	30°–60° [21] 40°–80° [23]
Angle of the axis of the indenters with respect to the direction of travel, ψ (see Fig. 6)	70°–90°	90° (all pins should contact the weld toe)
Working speed	3–5 mm/s	5–10 mm/s [21] 5–25 mm/s [23]
Other		The self-weight of the tool is sufficient [21, 23] Minimum of 5 passes [23]

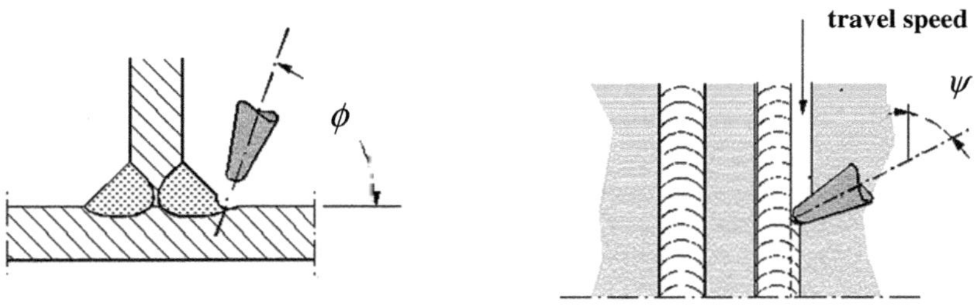

Fig. 6 Orientation of the HFMI tool with respect to the weld being treated

3.5 *Other Treatment Conditions*

Heat treatment and hot-dip galvanizing should not be performed after HFMI. HFMI introduces beneficial compressive residual stresses which may be reduced or eliminated by these operations. Fatigue strength of a HFMI treated component which is then treated by hot-dip galvanizing may have improved strength with respect to a hot-dip galvanized component without HFMI, where the annealing effects is diminishing the benefits of the treatment. In such a case, the fatigue strength improvement for HFMI treatment cannot be claimed and fatigue strength should be determined by fatigue testing.

Static local stresses near a weld toe are the results of both welding residual stresses and dead loads in structures. If the tensile residual stresses following welding are close to the yield strength of the material (as is normally assumed), the additional of a dead load will cause local yielding but will not result in increased maximum local stresses. HFMI treatment following the application of the dead load will produce compressive residual stresses in the critical weld toe region. On the other hand, if HFMI treatment is performed before the dead loads are applied; the compressive residual stresses following treatment may be partially counteracted by the local tensile stresses due to dead load. Thus, if significant dead loads are present on the structure during normal usage, it is recommended that the dead loads are applied prior to treatment, i.e. erect the structure with the welds untreated and then perform the treatment on-site [21].

4 Quality Control

Visual inspection of the HFMI groove following treatment consists of both qualitative and quantitative measures of the treated area. Due to differences in the HFMI tools and the wide variety of potential applications, certain details of proper treatment procedures and quantitative quality control measures are presented generally.

4.1 Qualitative Measures

Visual inspection following treatment includes an evaluation of the quality of the groove and the groove depth. The resulting groove must be smooth along all defined welds. A smooth and shiny groove without lines is one characteristic of a properly treated weld toe, see Fig. 5b. No thin line representing original fusion line should be visible in the groove. A thin crack-like line such as that shown in Fig. 7a is an indication that the weld fusion line has not been treated as previously described in Sect. 3.2. Dye penetrant or simple magnification with a ×3 to ×10 magnifying glass with proper surface illumination (minimum 350 lx) will be helpful in assessing the quality of the HFMI groove. Figure 7b shows an HFMI groove which is not smooth and shows indications of individual indenter strikes.

The HFMI groove must be continuous with no breaks. If the treatment cannot be per-formed without interruption, e.g. long welds or around corners, it is recommended that the operation be re-started at least 10 mm behind the stop position. No indications of undercut or porosity in the HFMI area can be visible. Similar qualitative measures have been specified by AASHTO [21].

HFMI produces significant local cold-forming of the material near the weld fusion line. If the indenters are directed excessively in one specific location, the resulting plastic displacement of the metal can result in a crack-like feature at the side of the HFMI groove as shown in Fig. 8. Failures of this type have been occasionally observed but not studied systematically [25]. The crack-like feature should be removed by light grinding and the weld toe should be retreated.

(a) (b)

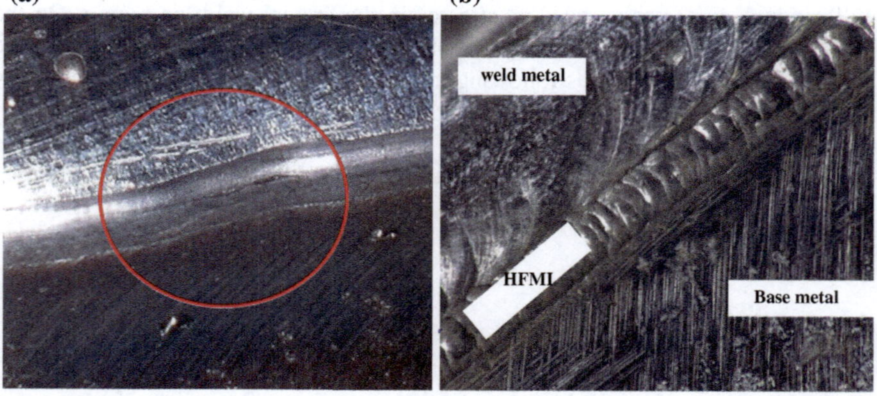

Fig. 7 The HFMI groove in **a** shows a thin crack-like defect which reduces or eliminates the effectiveness of the HFMI treatment [22], **b** shows a defect-free groove but with individual indenter strike still visible indicating the need for additional passes [24]

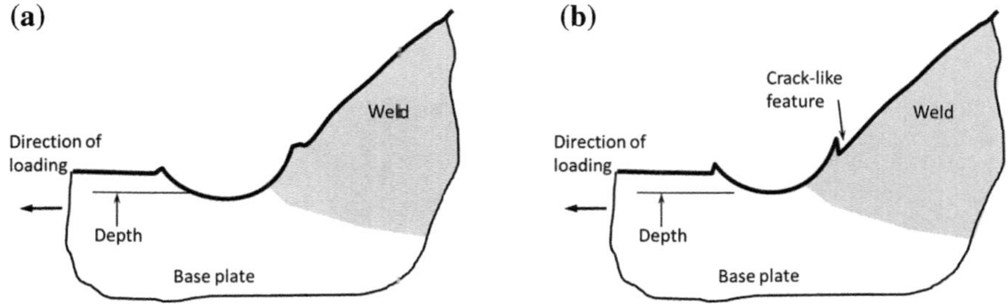

Fig. 8 **a** Proper profile of an HFMI groove which has no sharp or crack-like features and **b** improper HFMI groove profile which shows the presence of a crack-like feature due to plastic deformation of the material

4.2 Quantitative Measures

The depth of the groove is an excellent indicator of the extent of HFMI treatment [26]. Weld toe groove due to HFMI is generally not reducing fatigue strength, a minimum groove depth of about 0.1–0.2 mm is necessary to guarantee a complete weld toe treatment. Depending on the yield strength of the steel and the size of the indenters, typically the optimum HFMI groove will be 0.2–0.6 mm in depth and the width will be 3–6 mm, see Fig. 9 [21, 23, 27, 28]. However, it should be noted that no single groove dimension is optimal in all situations. A welded structure with relatively deep undercuts at the weld toe of which requires light grinding of the weld toe before HFMI will necessarily have deeper grooves following HFMI. Also, HFMI grooves in high strength steel structures will typically be shallower and narrower than grooves in low strength steel. Groove depth can be checked relatively easily by using simple depth gauges such as is shown in Fig. 10. Calipers can be used to measure the width of the groove. The center of the HFMI groove should

Fig. 9 The HFMI indentation depth following treatment should be 0.2–0.6 mm while the resulting width is typically 2–5 mm

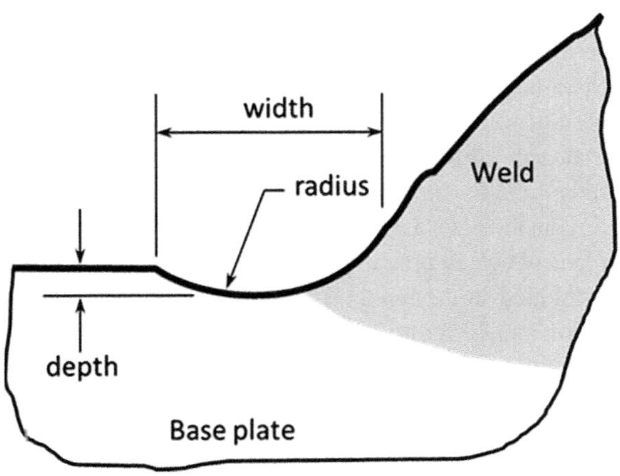

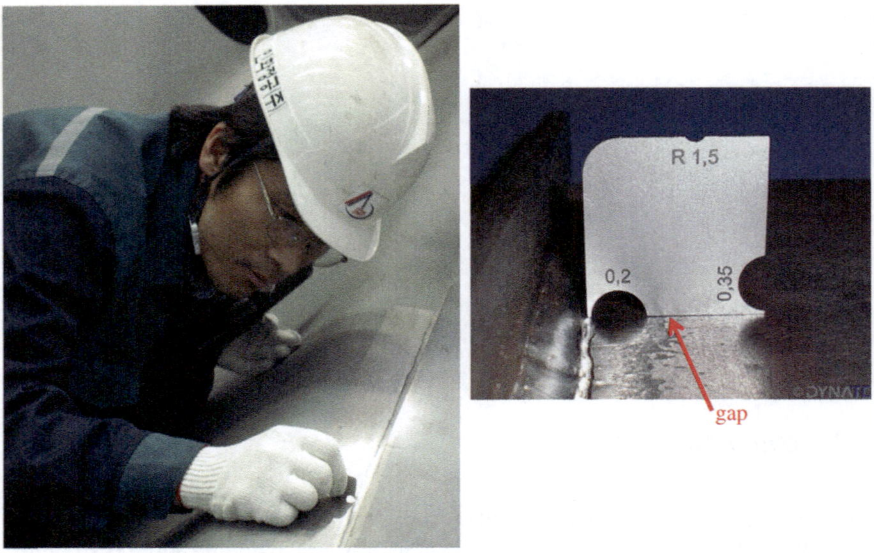

Fig. 10 Depth inspection using simple gauges [22]. The gap between the base plate and the gauge in the right-hand picture indicates that 0.2 mm has not been achieved

correspond to the fusion line of the weld. The portion of the HFMI groove in the weld metal must be between 25 and 75 % of the total HFMI groove width [27].

In large complex welded structures, welding heat input will not always be constant along a long weld. For this reason material hardness at the weld toe may vary and the HFMI treatment may need to be systematically adapted. HFMI groove dimension checks will be needed at regular intervals.

4.3 Documentation

A HFMI Procedure Specification (HFMI-PS) similar to a Welding Procedure Specification (WPS) should be prepared for the HFMI treatment. The HFMI-PS includes information concerning the component being treated; base and filler material; HFMI equipment type and power settings; number, size and shape of the indenters to be used; special inspection requirements including HFMI groove dimension, etc. Lopez Martinez and Haagensen have developed a HFMI-PS template which is prepared for each weld in a structure [29]. A HFMI-PS template is presented in Appendix 1. Similar, production data sheets, for treatment documentation can be found in Haagensen and Maddox [3] for other post weld improvement techniques, e.g. hammer-and needle peening.

4.4 Calibration

All of the available HFMI devices have variable power settings which can be adjusted de-pending on the material being treated and the indenter configuration. As a quality assurance measure, the intensity should be recorded in the HFMI-PS. In some cases, HFMI tool calibration is accomplished during treatment of a welded joint by ensuring that the resulting HFMI groove dimensions for a specified power setting and treatment time are consistent with pre-determined limits. The concept is similar to that used in the well-known Almen strip test which is common for measuring the intensity of shot peening and blasting operations. An example of calibration equipment for measurement of HFMI treatment intensity is shown in Fig. 11 [30]. Residual stress-free flat strips (200 mm × 20 mm × 4 mm) of S355J2 steel are held in a special fixture. HFMI is applied to the strip via the longitudinal slots. 4–5 passes with an HFMI tool with a predefined power setting are applied. Curvature of the strip, which is related to the resulting residual stress, is measured by means of a dial gauge. HFMI equipment producers have various calibration procedures and should be consulted for the specific tools used.

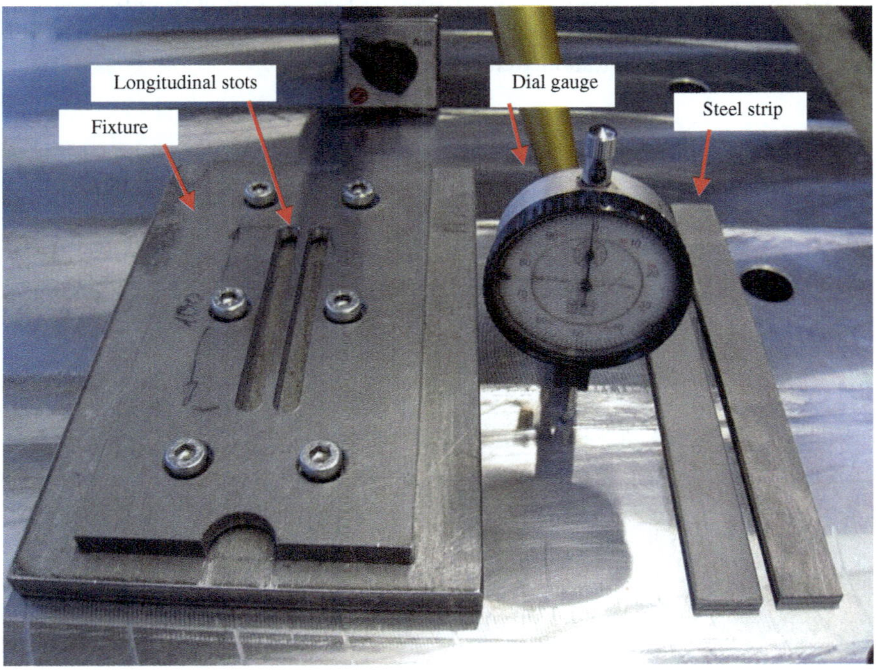

Fig. 11 Equipment needed to perform the Almen test-type calibration procedure [30]

5 Fatigue Strength Improvement Using HFMI

This guideline includes design resistance curves based on; nominal stress, structural hot-spot stress and effective notch stress assessment methods. These different assessment methods are detailed described in Hobbacher [31]. The HFMI improvement is applied to the weld toe and is intended to increase the fatigue lives of the weld treated from the view point of potential fatigue failure from the weld toe, examples of relevant weld details are shown in Fig. 12.

Possibility of a failure starting at some other location must always be considered. If the failure origin is merely shifted from the weld toe to the root there may be no significant improvement in fatigue life. It is emphasized that fatigue cracking from the root is governed by different design curves so toe treatment cannot be expected to provide any improvement in the general case. Improvement of details with incomplete penetration should be verified by fatigue testing or by analysis [31]. Examples of details in which root cracking might occur are shown in Fig. 13, but even nominally non-load-carrying welds may fail from the root when the toe has been improved. Consequently, when weld improvement is planned, full penetration welds or welds with extra-large throats should be used where possible, particularly for welds at the ends of cover plates and longitudinal stiffeners. In the IIW guidelines for the assessment of weld root fatigue [32] detailed procedures for various methods are outlined and could be used for assessing the life from the weld root side for partial penetration welds.

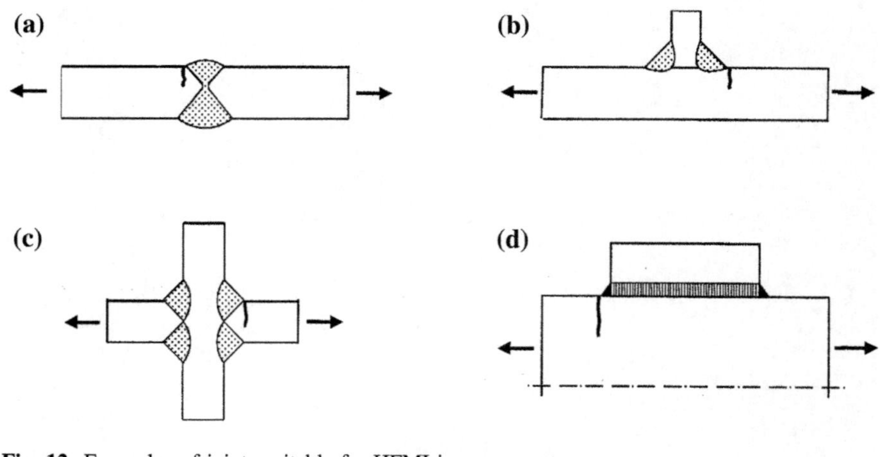

Fig. 12 Examples of joints suitable for HFMI improvement

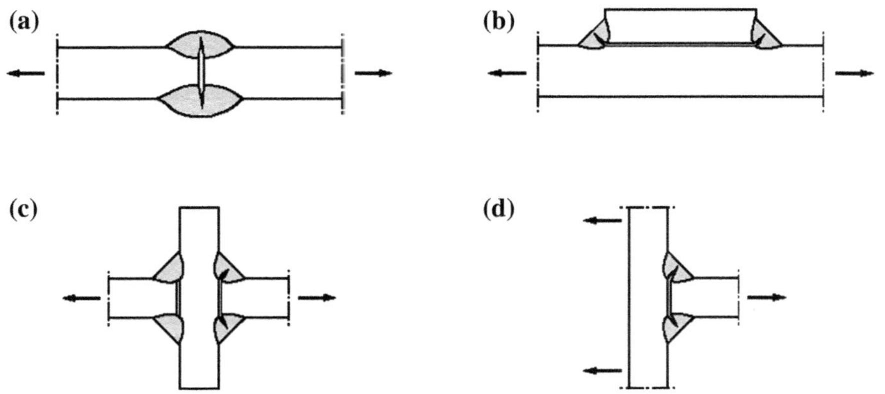

Fig. 13 Examples of details for which the improvement in fatigue strength is likely to be limited by the occurrence of root cracking. To ensure that a significant increase in fatigue strength is obtained full penetration welds should be used in case **a**, **c** and **d**, while an extra-large weld throat should be used in case **b**

5.1 *Fatigue Design Based on Nominal Stress*

All the fatigue design methods for HFMI improved welds (and also hammer-and needle peened welds [18]) are based on an assumed S-N slope of $m = 5$, and fatigue strength improvement factor are defined at $N = 2 \times 10^6$.

The benefit of HFMI treatment can be claimed only for details in design Class FAT 50 to FAT 90 in the IIW notation for S-N curves. This limitation is due to the fact that the higher classes include non-welded details, details whose lives are not governed by weld toe failure or welds that have been already been improved, e.g., a butt weld ground flush with the plate surface. Weld details lower than FAT 50 have not been studied experimentally with respect to HFMI improvement. For these details there is also increased risk of root side fatigue which is not influenced by HFMI. For steels with specified yield strength lower than 355 MPa, the proposed benefit consists of an upgrade by a factor of 1.6 applied to the stress range with an S-N slope of $m = 5$. S-N curves are shown in Fig. 14. However, the maximum class which can be claimed is the closest category below the FAT value obtained when the as-welded FAT value is multiplied by 1.6. For ease of computation, this corresponds to a four (4) fatigue class increase. For example, when a weld detail which, in the as-welded condition, would be classified as FAT 71 is treated by HFMI, the new FAT value is FAT 112. In Fig. 14 this S-N curve is denoted as 112 (71). For constant amplitude loading it is recommended that the slope parameter m is changed to 22 at 1×10^7 cycles. For more common case of variable amplitude loading the IIW Recommendations [31] use a S-N line that changes slope from m for $N < 1 \times 10^7$ to $m' = (2m - 1)$ for $N > 1 \times 10^7$. Thus, it is recommended to use $m' = (2m - 1) = 9$ for $N > 1 \times 10^7$. Because the S-N lines in the as-welded state follows a slope $m = 3$ and, after HFMI treatment, the slope is

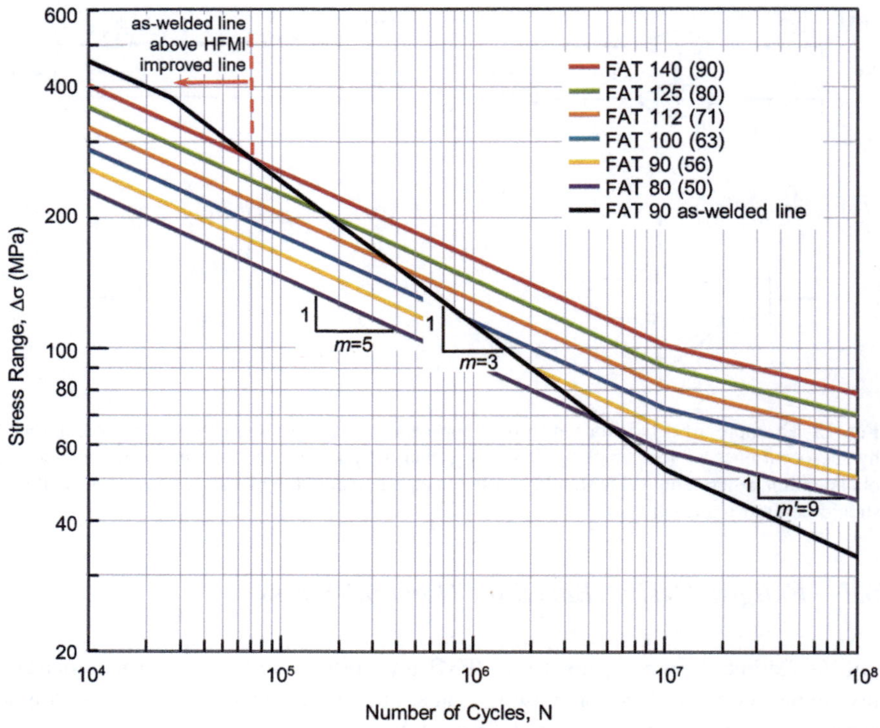

Fig. 14 Characteristic nominal stress S-N curves for HFMI improved welded joints in low strength steel ($f_y < 355$ MPa) for $R \leq 0.15$. The value in "()" represents the FAT class of the joint in the as-welded state according to Hobbacher [31]

changed to $m = 5$, the computed improvement in fatigue strength changes as a function of N. Consider the case of a welded joint in low strength steel ($f_y < 355$ MPa) which in the as-welded condition would be classified as FAT 90 ($m = 3$). HFMI treatment results in a FAT value of 140 ($m = 5$). The improved and as-welded S-N lines intersect at about N = 72 000 cycles. This is illustrated in Fig. 14 which shows that the FAT 90 and FAT 140(90) curves intersect at N = 72 000 cycles. Thus, for welded structures in low strength steel, fatigue strength improvement using HFMI will not be expected if the fatigue life is less than 72 000 cycles.

5.1.1 Thickness and Size Effects

The design proposal for HFMI is applied to plate thicknesses of 5–50 mm. Plate thickness and weld size influence the local stress concentration at the critical weld toe and the stress gradient through the plate thickness, therefore, both the nominal stress assessment method and the structural hot spot stress method require a

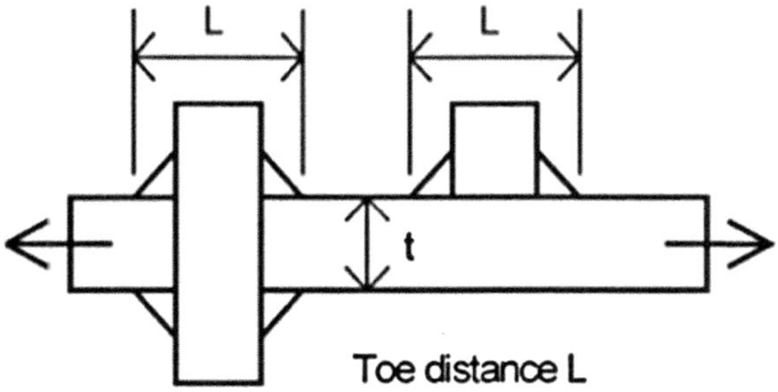

Fig. 15 Definition of L used to determine the thickness correction factor [31]

thickness reduction factor for plate thicknesses exceeding 25 mm. It is proposed to use the same factors which occur in the existing IIW guideline, Hobbacher [31]. The thickness correction factor $f(t)$ for toe ground joints is given in Eq. (1).

$$f(t) = \left(\frac{25}{t_{eff}}\right)^{0.2} \tag{1}$$

where t_{eff} = L/2 for $L/t < 2$ and $t_{eff} = t$ for $L/t \geq 2$. As seen in Fig. 15, L is the sum of the thickness or length of the attachment and the weld leg lengths.

5.1.2 Influence of Steel Strength

One fatigue class increase in strength (about 12.5 %) for every 200 MPa increase in static yield strength is proposed and shown to be conservative with respect to all available data [33].

The recommendation for design includes a four (4) fatigue class increase in strength for joints fabricated from $f_y < 355$ MPa steel with respect to the nominal fatigue class in the as-welded condition. One additional fatigue class increase in fatigue strength for every 200 MPa increase in static yield strength is proposed. The specific fatigue class increase is defined for $N = 2 \times 10^6$ cycles and assumes an S-N slope $m = 5$ for HFMI treated welds and $m = 3$ for welds in the as-welded state. Figure 16 shows a proposed increase in number of FAT classes as a function of yield strength.

The characteristic nominal stress S-N curves for HFMI improved welded joints in low strength steel ($f_y < 355$ MPa) is shown in Fig. 13. The characteristic nominal stress S-N curves for HFMI improved welded joints in steel strength (355 MPa $\leq f_y <$ 550 MPa) is shown in Fig. 17, in steel strength (550 MPa $\leq f_y <$ 750 MPa) in Fig. 18, in steel strength (750 MPa $\leq f_y <$ 950 MPa) in Fig. 19 and in steel strength ($f_y \geq$ 950 MPa) in Fig. 20.

A HSS welded joint classified as FAT 80 ($m = 3$) in the as-welded condition would be FAT 180 ($m = 5$) following HFMI treatment. A welded joint classified as FAT 90 ($m = 3$) in the as-welded condition would similarly be FAT 180 ($m = 5$) following HFMI treatment. This is due to the restriction that the highest S-N curve

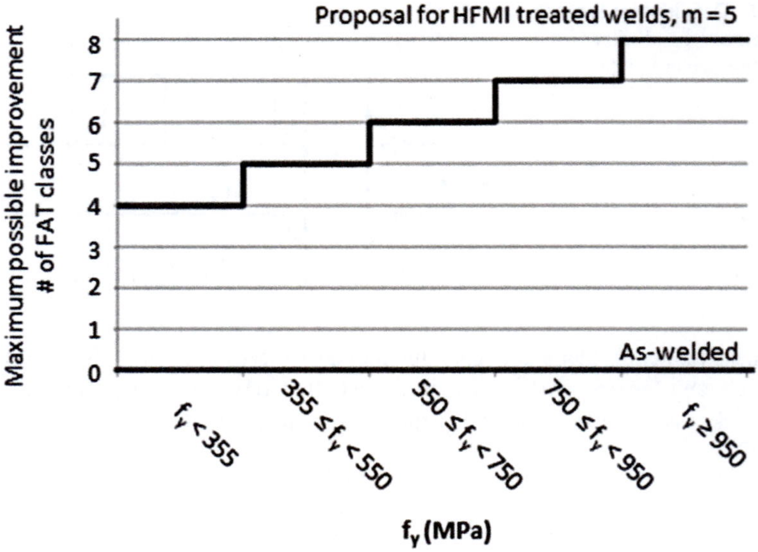

Fig. 16 Maximum increase in the number of FAT classes as a function of f_y [33]

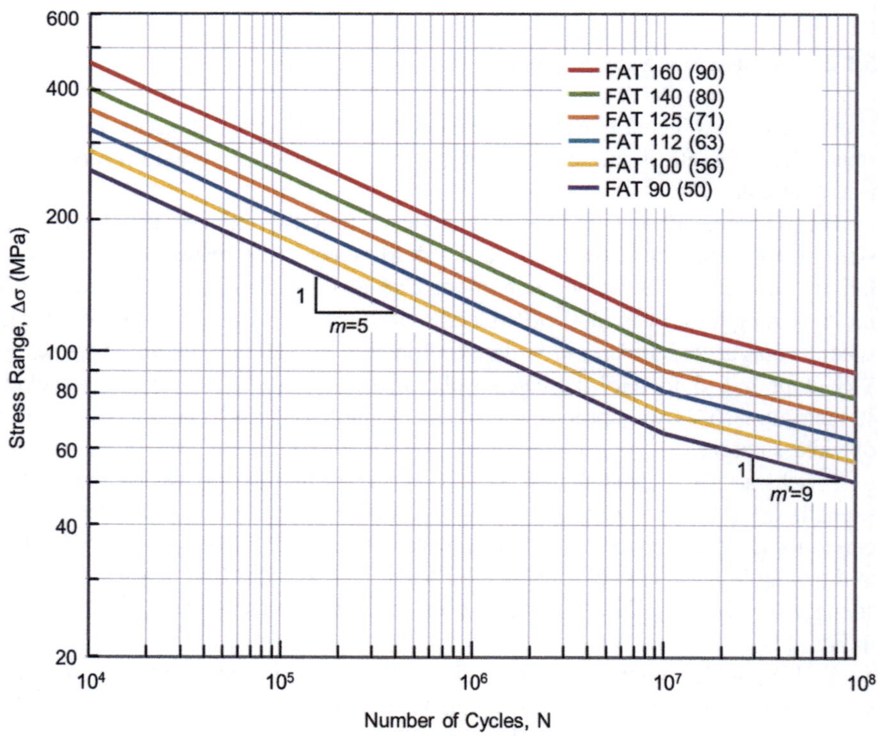

Fig. 17 Characteristic nominal stress S-N curves for HFMI improved welded joints for high strength steels, 355 MPa $\le f_y < 550$ (MPa) for R $\le$ 0.15. The value in "()" represents the FAT class of the joint in the as-welded state according to Hobbacher [31]

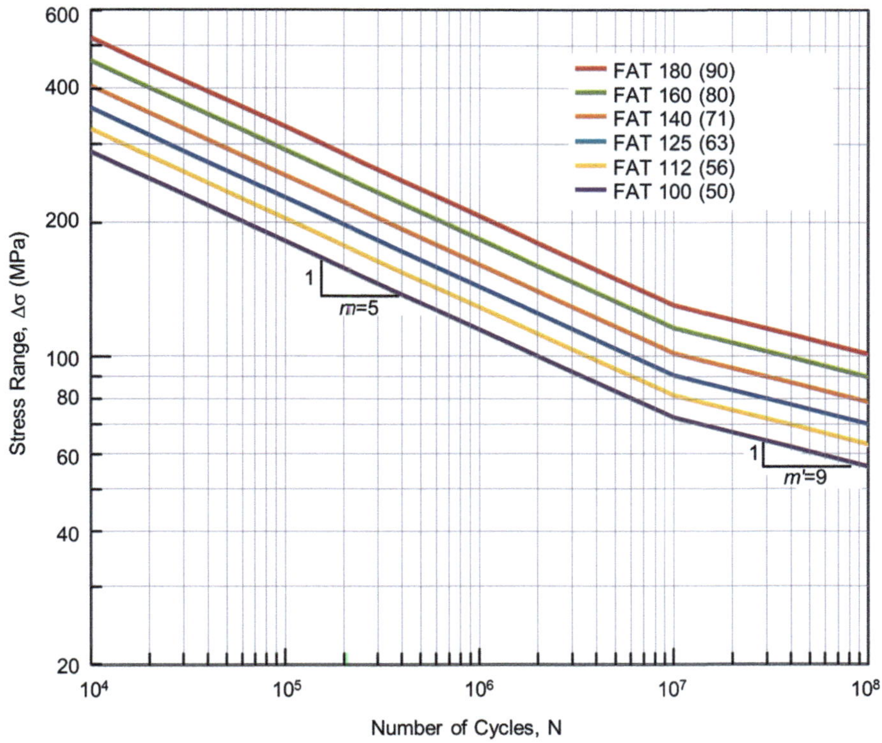

Fig. 18 Characteristic nominal stress S-N curves for HFMI improved welded joints for high strength steels, 550 MPa $\leq f_y < 750$ (MPa) for R $\leq$ 0.15. The value in "()" represents the FAT class of the joint in the as-welded state according to Hobbacher [31]

that can be claimed following HFMI improvement is FAT 180 ($m = 5$). This restriction is based on the current lack of available experiment data.

The change in slope of the S-N lines between the as-welded state and following HFMI treatment results in a change in the computed fatigue strength improvement as a function of N. For a low strength steel welded detail it was shown that the S-N lines intersect at about N = 72 000 cycles. For higher strength steels the computed cycle limits below which HFMI does not provide benefit are shown in Table 3. As can be seen from this table, for steels with yield strength $f_y \geq 750$ MPa, it is computed that HFMI would have a beneficial effect even into the low cycle fatigue regime, N < 10000 cycles.

5.1.3 Loading Effects

In the IIW recommendations for fatigue design of welded components in the as-welded state, the nominal normal stress range is assumed to remain below 1.5 * fy

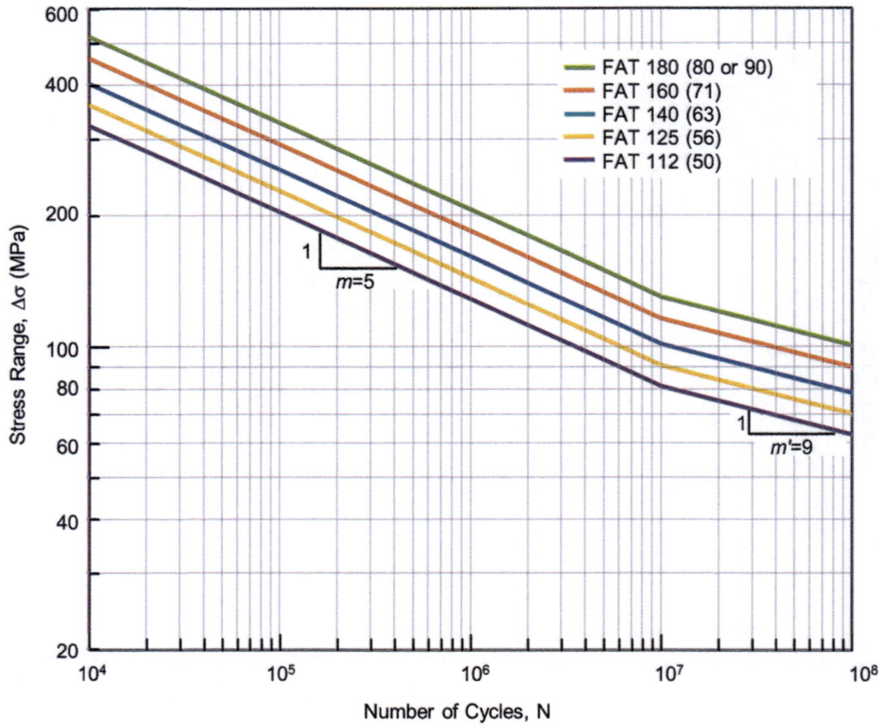

Fig. 19 Characteristic nominal stress S-N curves for HFMI improved welded joints for high strength steels, 750 MPa $\leq f_y < 950$ (MPa) for R $\leq$ 0.15. The value in "()" represents the FAT class of the joint in the as-welded state according to Hobbacher [31]

while the nominal shear stress range must remain below 1.5 * $f_y/\sqrt{3}$ [31]. In the structural hot spot stress approach hot spot stress range is assumed to remain below 2 * f_y [32]. For welded structures improved by and needle peening or hammer peeing, the techniques are not suitable for R > 0.5 or when $\sigma_{max} > 0.8\ f_y$ [3]. These restrictions are intended to influence the fatigue design assessment of structures when the beneficial compressive residual stress state due to post weld improvement may not be stable.

5.1.4 Stress Ratio

HFMI treated welds can have up to 8 FAT classes of improvement depending on the material strength, welded joint geometry, etc. The stress ratio influence is expressed as a penalty with respect to the maximum increase in the number of FAT classes as a function of f_y. These are given in Table 4.

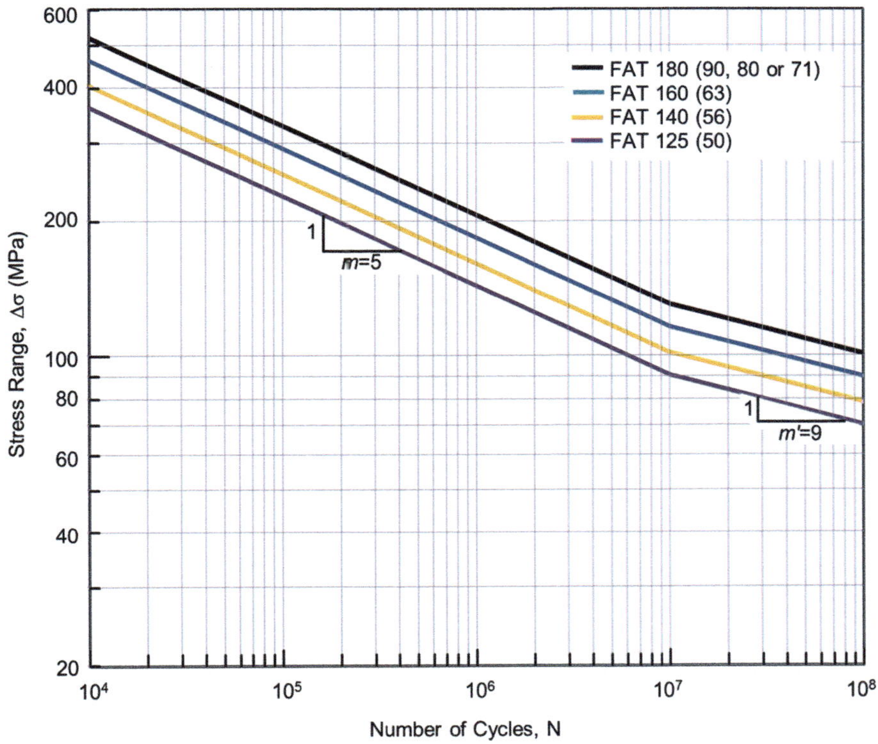

Fig. 20 Characteristic nominal stress S-N curves for HFMI improved welded joints for high strength steels, 950 MPa $\leq f_y$ for $R \leq 0.15$. The value in "()" represents the FAT class of the joint in the as-welded state according to Hobbacher [31]

Table 3 Computed cycle limit below which HFMI is not expected to result in fatigue strength improvement as a function of steel strength

f_y (MPa)	N (cycles)
$f_y < 355$	72,000
$355 \leq f_y < 550$	30,000
$550 \leq f_y < 750$	12,500
≥ 750	<10,000

Table 4 Minimum reduction in the number of FAT classes in fatigue strength improvement for HFMI treated welded joints based on R ratio

R ratio	Minimum FAT class reduction
$R \leq 0.15$	No reduction due to stress ratio
$0.15 < R \leq 0.28$	One FAT class reduction
$0.28 < R \leq 0.4$	Two FAT classes reduction
$0.4 < R \leq 0.52$	Three FAT classes reduction
$0.52 < R$	No data available. The degree of improvement must be confirmed by testing

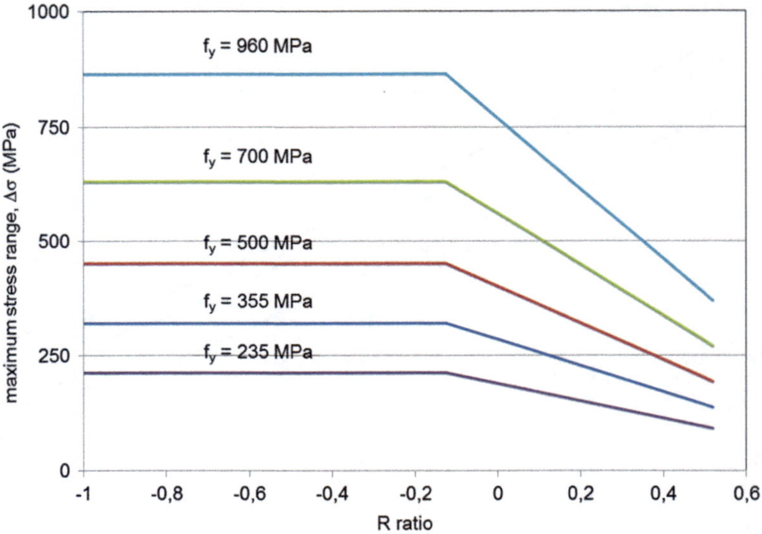

Fig. 21 Limitation on maximum constant amplitude stress range, $\Delta\sigma$, that can be applied to a weld in order to claim benefit from HFMI treatment (in MPa)

While it has not been studied systematically, the previously mentioned limitation that $\sigma_{max} \leq 0.8 \cdot f_y$ should apply also to HFMI treated welded joints. This information is summarized in Fig. 21 which shows the limitation on maximum constant amplitude stress range, $\Delta\sigma$, as a function of R for several values of f_y. In the case $R < -0.125$ loading, the horizontal lines correspond to $\Delta\sigma = 0.9 \cdot f_y$. If $\Delta\sigma$ exceed the stress range values in this figure, benefit from HFMI treatment cannot be claimed without supporting tests. This table illustrates the benefit of HFMI especially for the higher yield strength materials. For low f_y materials with high R, HFMI can improved the so-called fatigue limit but it is not expected to improve fatigue strength in the finite life regime. A computation example involving high R-ratio fatigue loading is given in Appendix 2.

5.1.5 Variable Amplitude Loading

In many design situations it is convenient to transform the variable amplitude history into an equivalent constant amplitude value based on rainflow counting and Eq. (2) [34].

$$\Delta\sigma_{eq} = \left(\frac{1}{D} \cdot \frac{\sum \Delta\sigma_i^m N_i + \Delta\sigma_k^{(m-m')} \cdot \sum \Delta\sigma_j^{m'} N_j}{\sum N_i + \sum N_j} \right)^{\frac{1}{m}} \quad (2)$$

In this equation $\Delta\sigma_k$ is the stress range associated with the knee computed at N = 1 × 10^7, N_i is the number of cycles of stress $\Delta\sigma_i$ where $\Delta\sigma_i > \Delta\sigma_k$, N_j is the number of cycles of stress $\Delta\sigma_j$ where $\Delta\sigma_i < \Delta\sigma_k$, m is the slope of S-N line above the knee point, $m' = 2m - 1$ is the slope of S-N line below the knee point and D is the damage sum, e.g., $D = 0.5$. As previously discussed $m = 5$ and $m' = 9$ are proposed for HFMI treated welds. Yildirim and Marquis [17] showed that Eq. (2) can be used to correlated variable amplitude and constant amplitude fatigue data for welds improved using HFMI.

In the case of HFMI treated welds, a significant portion of the fatigue strength improvement is attributed to the presence of compressive residual stresses. Therefore, any change in the residual stress state which might occur during variable amplitude loading could have significant influence on the resulting fatigue strength. For example, it has been observed that, for HFMI treated high strength steel welds, the failure mode changes depended on whether the fatigue loading is constant or variable amplitude [25]. Variable amplitude loading using a load history which contained some large stress range cycles led to failure in the HFMI groove while constant amplitude loading resulted in other failure locations.

The information contained in Fig. 21 can also be used to assess whether the fatigue design guidelines outlined in this proposal can be used for an HFMI treated welded joint subjected to variable amplitude loading. The stress amplitude and R ratio in the loading spectrum can be compared with Fig. 21. If the largest cycle exceeds the limit value for a given f_y, then the benefit of HFMI cannot be claimed without fatigue testing. A computation example involving variable amplitude loading is given in Appendix 2.

5.2 Fatigue Design Based on Structural Hot-Spot Stress

For fatigue design based on the structural hot spot stress, stress analysis procedures as described by Niemi et al. are recommended [35]. As is the case with the nominal stress method, fatigue resistance curves for HFMI improved welds are based on an assumed hot spot S-N slope of $m = 5$ in the region $1 \times 10^4 \leq N < 1 \times 10^7$ cycles and $m' = 9$ for $1 \times 10^7 \leq N$. Characteristic curves are defined at $N = 2 \times 10^6$.

In the case of steel in the as-welded condition, two structural hot spot stress design curves are proposed. For load-carrying fillet welds a FAT 90 curve is recommended and for non-load carrying welds a FAT 100 curve is recommended. For HFMI improved welds the corresponding S-N characteristic curves are given in Table 5 for different steel grades.

In the context of the nominal stress based approach to fatigue design of HFMI treated welds, it has previously been noted that the highest allowable S-N curve corresponds to FAT 180. When the structural hot spot stress method is used a computation problem may arise for welded details with relatively small structural stress concentrations, i.e., the hot spot stress system must be limited so as not to result in a S-N curve greater than FAT 180 in the nominal stress system.

Table 5 Existing IIW FAT classes for structural hot spot approach for as-welded and improved joints and the proposed FAT classes for HFMI treated joints as a function of f_y

	Load-carrying fillet welds		Non-load carrying fillet welds	
f_y (MPa)	FAT	$K_{S,min}$	FAT	$K_{S,min}$
	As-welded, m = 3 [31]			
All f_y	90	–	100	–
	Improved by HFMI, $m = 5$			
<355	140	–	160	–
355–550	160	–	180	–
550–750	180	–	200	1.15
750–950	200	1.15	225	1.25
>950	225	1.25	250	1.40

Minimum structural hot spot stress concentrations for HFMI improved welded connections are also shown

By defining the structural hot spot stress concentration as the ratio of structural stress to nominal stress

$$K_S = \frac{\sigma_S}{\sigma_{nom}} \tag{3}$$

the minimum K_S for HSS should be limited according to the values given in Table 5.

As with the nominal, the fatigue strength thickness correction factor in Eq. (1), fatigue class reduction due to stress ratio, and design guidelines for variable amplitude loading should be used in conjunction with the hot spot stress approach.

5.3 *Fatigue Design Based on Effective Notch Stress*

In 2008 the IIW approved a guideline encompassing fatigue design recommendations based on the effective notch stress approach to fatigue assessment [31]. The design proposal is to perform the stress analysis using an artificial notch radius $\rho_f = 1$ mm using procedures as described by Fricke [36]. Fatigue resistance curves for HFMI improved welds are based on an assumed S-N slope of $m = 5$ in the region $1 \times 10^4 \leq N < 1 \times 10^7$ cycles and $m' = 9$ for $1 \times 10^7 \leq N$. As is the case with the nominal stress method, characteristic curves are defined at $N = 2 \times 10^6$. For HFMI improved welds the corresponding S-N characteristic curves are given in Table 6 for different steel grades.

In the context of the nominal stress based approach to fatigue design of HFMI treated welds, it has previously been noted that the highest allowable S-N curve corresponds to FAT 180 $m = 5$. When the effective notch stress method is used a computation problem may arise for welded details with relatively small structural

Table 6 Existing IIW characteristic curve based on the effective notch stress approach for as-welded joints and proposed characteristic curves based on the effective notch stress approach for HFMI improved joints as a function of f_y

f_y (MPa)	Effective notch stress characteristic curve modelled using $\rho_f = 1$ mm
	As-welded, $m = 3$ [31].
All f_y	225
	Improved by HFMI, $m = 5$
$235 < f_y \leq 355$	320
$355 < f_y \leq 550$	360
$550 < f_y \leq 750$	400
$750 < f_y \leq 950$	450
$950 \leq f_y$	500

stress concentrations, i.e., the effective notch stress system must be limited so as not to achieve an S-N curve greater than FAT 180 in the nominal stress system. Fricke [36] defines the effective notch stress concentration as the ratio of effective notch stress to structural stress.

$$K_W = \frac{\sigma_W}{\sigma_S} \tag{4}$$

Fricke proposes that a minimum K_w of at least 1.6 should be used for low stress concentration details. This means that both the structural hot-spot stress and the effective notch stress at a weld toe need to be checked. If it the computed K_w based on finite element calculations with $\rho_f = 1$ mm is less than 1.6, then a minimum values of $K_w = 1.6$ should be used for fatigue assessment. No additional requirements are needed for HFMI treated welds because K_w is defined with respect to σ_S. It is sufficient to guarantee that the minimum K_S for HSS should be limited according to the values given in Table 5.

The fatigue strength recommendations in Table 6 are based on the assumption that $\Delta\sigma$ is computed in terms of the maximum principal stress range in the notch. If von Mises stress is used, S-N curves must be reduced by one fatigue class.

As with the nominal, the fatigue class reduction due to stress ratio and design guidelines for variable amplitude loading should be used in conjunction with the effective notch stress approach. As a highly local method, effective notch analysis does not require a fatigue strength thickness correction factor.

Appendix 1: HFMI Procedure Specification

Used for each treated weld in the structure as a quality assurance measure (Table 7).

Table 7 HFMI Procedure Specification

HFMI procedure specification for weld toe improvement				
Welding specification		**Weld joint identification**		
Base material		Type		
Thickness		Location		
Filler material		Identification		
Welding procedure no.		UPPS-number		
		Rev.		
		Date		
Equipment		**Photo of treated welded joint**		
Make and Model				
Power (kW)				
Tip diameter (mm)				
Weight (kg)				
Impact frequency				
Impact amplitude				
Ultrasonic frequency				
Treatment data				
Position				
Work angle side				
Work angle ahead				
Travel speed				
Number of passes				
Treated length				
Time of treatment				
Tool changes				
Cause				
Operator				
Name				
Experience (h)				
Treated length				
Date of treatment		**Remarks**		
Inspection				
Visual				
Photo				
Measurement				
Equipment				
Results				
Toe radius (mm)				
Weld angle (deg)				
Groove depth (mm)				
Groove width (mm)				

(continued)

Table 7 (continued)

HFMI procedure specification for weld toe improvement			
Approvals			
	Contractor	**Client**	**Survey authority**
Name			
Date			
Responsible			
Signature			

Appendix 2: Design Examples

Example 1: Nominal stress design for a detail subjected to high R ratio

Example: Consider the case of a welded detail which in the as-welded condition corresponds to FAT 63. The joint is fabricated from f_y = 960 MPa steel and will be subjected to R = 0.5 loading.

Question: What is the suitable characteristic line for design?

Solution: Based on Fig. 16, an increase of eight fatigue classes can be claimed for R $\leq$ 0.15. The resulting S-N curve FAT 160 (63) is shown in Fig. 20. Based on Table 4 the S-N curve for R = 0.5 is reduced by three fatigue classes with respect to R $\leq$ 0.15 loading so the characteristic curve is considered to represent five FAT class improvement, i.e., FAT 112. This is shown in Fig. 22. With respect to Fig. 21, the limitation on maximum stress range that can be applied to a weld in order to claim benefit from HFMI treatment in this example is $\Delta\sigma$ = 340 MPa. This value would correspond to N $\approx$ 7760 cycles. The characteristic line is shown in Fig. 22 is therefore considered to be valid over the entire life range shown.

Example 2: Nominal stress design for a detail subjected to variable amplitude loading

Example: Consider the case of a longitudinal welded attachment from steel with f_y = 700 MPa, subject to variable amplitude loading. The target fatigue life is 1×10^7 cycles. Each load cycle has R = 0 and $\Delta\sigma_{eq} = 0.20\ \Delta\sigma_{max}$ based on Eq. (2).

Question: Will HFMI be an effective improvement technology for this design case?

Solution: In the as-welded condition a typical longitudinal attachment is FAT 71. According to Fig. 16, a detail fabricated from f_y = 700 MPa shows six (6) fatigue classes improvement due to HFMI. The resulting characteristic S-N curve is FAT 140 (71), see Fig. 18.

This characteristic curve intersects 1×10^7 cycles at $\Delta\sigma_{eq}$ = 102 MPa. Based on the design load spectrum the maximum stress which occurs is 102/0.20 = 510 MPa. According to Fig. 21, the maximum allowable stress range for f_y = 700 MPa at R = 0 is 560 MPa. Thus, HFMI is expected to be fully effective for this component.

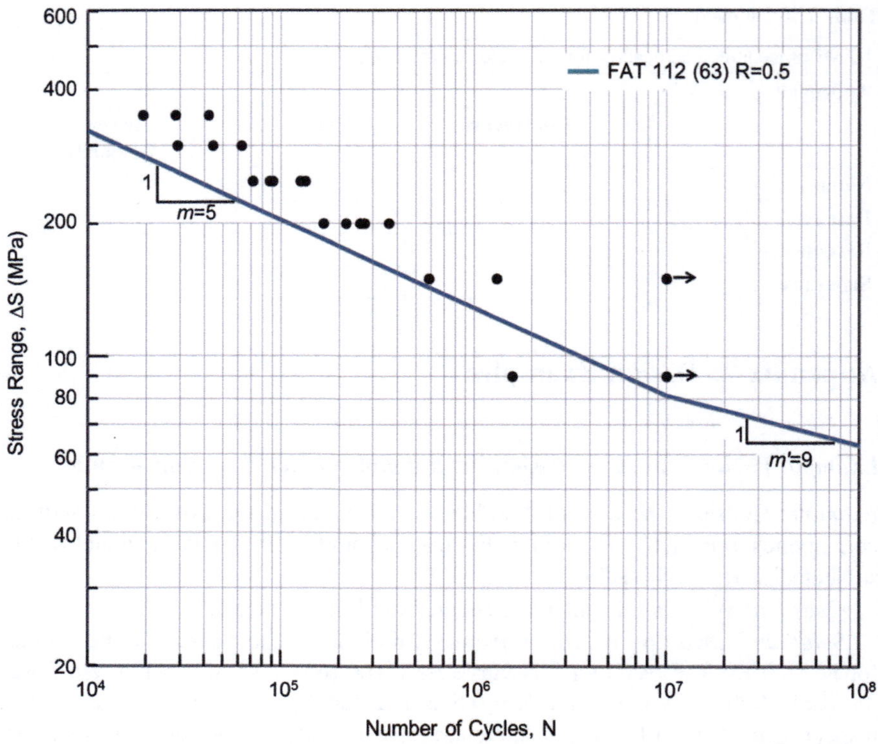

Fig. 22 Example of the characteristic nominal stress S-N curves for HFMI improved welded joints for high strength steels f_y = 960 MPa for R $\leq$ 0.5. The as-welded joint is FAT 63; for the material strength there are +8 classes while for R ratio there are −3 classes resulting in a total of 5 classes increase, i.e., FAT 112

For comparison purposes, the characteristic fatigue life for a FAT 71 welded joint subjected to $\Delta\sigma_{eq}$ = 102 MPa would have a characteristic fatigue life of N_f = 675 000 cycles. Therefore, for this design case, HFMI is computed to result in a fatigue life increase of 14.8×.

Example 3: Structural hot spot based assessment for a detail subjected to variable amplitude loading

Example: Consider an HFMI treated non-load carrying structural detail which is subjected to variable amplitude loading for which the cycle range distribution is approximately log-linear. Assume that $\Delta\sigma_{eq} = 0.387\ \Delta\sigma_{max}$ when computed according to Eq. (3) with D = 0.5. Each cycle has R = −1. The structure is fabricated from S960 steel. The computed structural stress concentration is K_s = 1.21.

Question: Construct the characteristic line and compare to experimental results.

Solution: As shown in Table 5, a non-load carrying detail fabricated from S960 (f_y = 960 MPa) steel treated by HFMI has a resulting structural hot spot S-N characteristic curve of FAT 250 with $K_{s,min}$ = 1.4. Because the computed

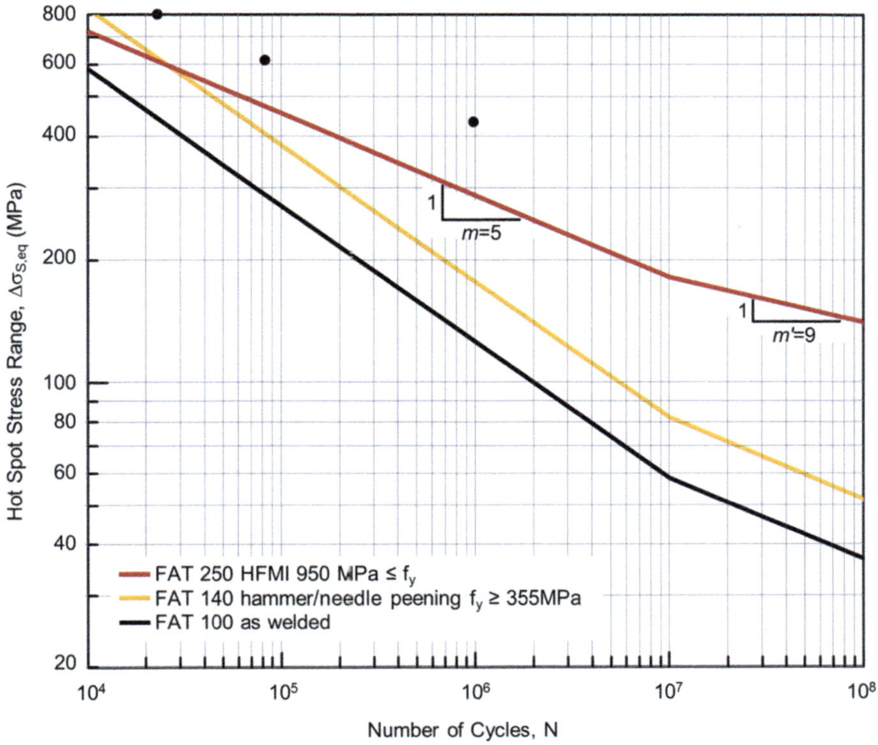

Fig. 23 Example of the characteristic structural hot-spot stress S-N curves for HFMI improved welded joints for high strength steels f_y = 960 MPa for R $\leq$ 0.15

Table 8 Experimental data for a S960 steel weld treated by HFMI

σ_{max} (MPa)	N_f
733	22,900
567	82,100
400	978,000

$K_s < K_{s,min}$, $K_{s,min}$ is used to compute σ_s. The loading is R = −1 so $\Delta\sigma_{s,eq}$ = (0.387 × 2)·$K_{s,min}$·σ_{max}. Results are shown in Fig. 23. From the data table, Table 8, it can be seen that the maximum nominal stress range for the first two experimental points exceed the limiting curves in Fig. 21. Therefore, the FAT 250 line is *not necessarily* expected to be conservative with these two points since the beneficial residual stresses may relax out. A conservative approach would be to assume that joints with such high nominal stresses behave according to the as-welded line. Alternatively, the effective notch method with a FAT 225 S-N curve could be using $\rho_f = \rho + 1$ mm with ρ as the actual HFMI groove radius.

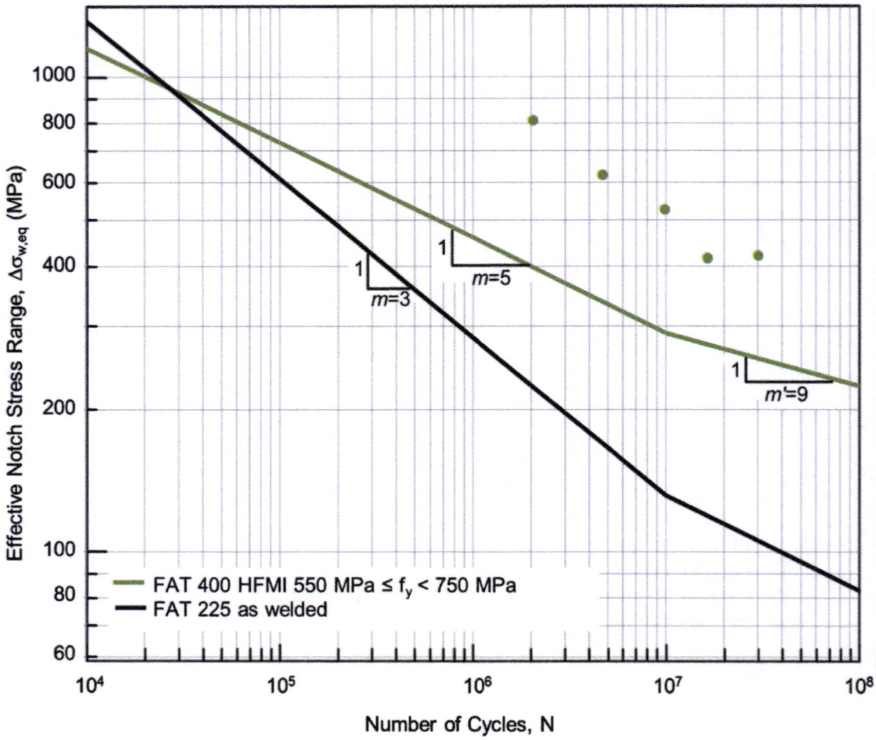

Fig. 24 Example of the characteristic effective notch stress S-N curves for HFMI improved welded joints for high strength steels f_y = 700 MPa for R $\leq$ 0.15

Example 4: Effective notch stress based assessment for a detail subjected to variable amplitude loading

Example: Consider an HFMI treated structural detail which is subjected to variable amplitude loading for which the cycle range distribution is approximately Gaussian. Assume that $\Delta\sigma_{eq} = 0.504 \cdot \Delta\sigma_{max}$ when computed according to Eq. (3) with D = 0.5. Each cycle has R = −1. The structure is fabricated from S700 steel.

Question: Construct the characteristic line and compare to experimental results.

Solution: As shown in Table 6, a detail fabricated from S700 (f_y = 700 MPa) steel treated by HFMI has a resulting effective notch method characteristic S-N curve of FAT 400. Because the loading is R = −1, $\Delta\sigma_{w,\,eq} = (0.504 \times 2) \cdot \sigma_{max}$. If it is assumed that K_s = 1.22 and K_w = 2.08, the maximum nominal stress range for all of the experimental points would be below the limiting curves in Fig. 21. Therefore, the FAT 400 is expected to be conservative with respect to 95 % of all the experimental data. Results are shown in Fig. 24. It is clear that the experimental data, Table 9, is conservative with respect to the characteristic line.

Table 9 Experimental data for a S960 steel weld treated by HFMI

$\sigma_{w, max}$ (MPa)	N_f
417	16,310,000
421	30 010,000
526	9,813,000
622	4,666,000
810	2,052,000

References

1. Marquis G. and Barsoum Z., Fatigue strength improvement of steel structures by HFMI: Proposed procedures and quality assurance guidelines, Welding in the World, No. 58, pp. 19–28, 2014
2. Marquis G., Mikkola E., H. Yildirim and Barsoum Z., Fatigue strength improvement of steel structures by HFMI: Proposed fatigue assessment guidelines, Welding in the World, No. 57, pp. 803–822, 2013
3. Haagensen P.J. and Maddox S.J., IIW Recommendations on methods for improving the fatigue strength of welded joints, ISBN:9781782420644, Woodhead Publishing, 2013
4. Barsoum Z. and Jonsson B., Influence of weld quality on the fatigue strength in seam welds, Engineering Failure Analysis, No. 18, pp. 971–979, 2011
5. Holmstrand T., Mrdjanov N., Barsoum Z., Åstrand E., Fatigue life assessment of improved joints welded with alternative welding techniques, Engineering Failure Analysis, pp. 10–21, Volume 42, 2014
6. Volvo Group weld quality standard, STD 181-0004; 2008
7. Jonsson B., Samuelsson J., Marquis G., Development of weld quality criteria based on fatigue performance, Welding in the World, Volume 55, Issue 1112, pp 79–88, 2011
8. Khurshid M., Barsoum Z. and Marquis G., Behavior of compressive residual stresses in high strength steel welds induced by High Frequency Mechanical Impact treatment, ASME Journal of Pressure Vessel Technology, pp. 1–8, Vol. 136, 2014
9. Applied Ultrasonics. In: http://www.appliedultrasonics.com/.
10. Structural Integrity Technologies Inc. (SINTEC) In: http://sintec.ca.
11. Lets Global. In: http://www.lets-global.com/.
12. Huo, L., Wang, D., Zhang, Y., Investigation of the fatigue behavior of the welded joints treated by TIG dressing and ultrasonic peening under variable-amplitude load International Journal of Fatigue, vol. 27, pp. 95–101, 2005
13. Pfeifer. In: http://www.pfeifer.de/.
14. Pitec. In: http://www.pitec-gmbh.com/.
15. Sonats. In: http://www.sonats-et.com/.
16. Bousseau, M., Millot T. Fatigue life improvement of welded strucrtures by UNP compared to TIG dressing, International Institute of Welding., Paris, Document XIII-2125-06, 2006
17. Yildirim, H.C. and Marquis, G.B., A Round Robin study of high frequency mechanical impact (HFMI)-treated welded joints subjected to variable amplitude loading, Welding in the World, Volume 57, Issue 3, pp 437–447, 2013.
18. Yildirim, H.C. and Marquis, G.B., Overview of Fatigue data for high frequency mechanical impact treated welded joints, Welding in the World, 57, issue 7/8, 2012.
19. ISO 5817:2006; Welding–Fusion-welded joints in steel, nickel, titanium and their alloys (beam welding excluded)–Quality levels for imperfections, 2006.
20. Yekta R.T., Ghahremani K and Walbridge S, Effect of quality control parameter variations on the fatigue performance of ultrasonic impact treated welds, Int J Fatigue, 55, 2013, pp. 245–256.

21. AASHTO LRFD bridge construction specifications Chapter 11.9 ultrasonic impact treatment, 3rd Ed, American association of state highway and transportation officials: LFRD bridge design specifications, Washington, D.C., 2010.
22. Neher, M., HiFIT. Presentation to the IIW Commission XIII intermediate meeting, Espoo, Finland 15-16 March, 2012.
23. Applied Ultrasonics, Esonix UIT application guide: Post weld treatment for fatigue enhancement carbon steel welded structures. Alabama, USA, 2006.
24. Le Quilliec, G., Lieurade, H.-P., Bousseau, M., Drissi-Habti, M., Inglebert, G., Macquet, P. and Jubin, L., Fatigue behaviour of welded joints treated by high frequency hammering: Part 1, Experimental study International Institute of Welding, Paris, IIW Document XIII-2394-11, 2011.
25. Marquis, G.: Failure modes and fatigue strength of improved HSS welds, Engineering Fracture Mechanics., Vol.77, No.11, pp. 2051−2062, 2010.
26. Yekta, R. T., and Walbridge, S., Acceptance Criteria for Ultrasonic Impact Treatment (UIT), Ontario Ministry of Transportation, Report HIIFP-110, St. Catherines, Ontario, Canada, 2012.
27. Tilly, G. P., Jackson, P. A., Maddox, S. J. and Henderson, R., Proceedings of the ICE - Bridge Engineering, Volume 163, Issue 3, 01 September 2010, pp. 147–152.
28. Y. Kudryavtsev and J. Kleiman: Measurement of Residual Stresses in Welded Elements and Structures by Ultrasonic Method. International Institute of Welding. IIW Document XIII-2339-10, 2010.
29. Lopez Martinez, L. and Haagensen, P. J., Life extension of Class F and Class F2 details using ultrasonic peening International Institute of Welding, Paris, IIW Document XIII-2143-06, 2006.
30. PIT 10 Almen test: PIT equipment calibration procedures, PITEC GmbH, Heudorf, Germany, p. 6, 2011.
31. Hobbacher, A., IIW Recommendations for Fatigue Design of Welded Joints and Components, WRC, New York, 2009.
32. Fricke W. IIW guideline for the assessment of weld root fatigue, Welding in the World, 57, pp. 753−791, 2013.
33. Yildirim, H. C. and Marquis, G. B., Fatigue strength improvement factors for high strength steel welded joints treated by high frequency mechanical impact. Int J Fatigue, 44, 2012, pp. 168−176.
34. Niemi, E., Random loading behavior of welded components, in Proc. of the IIW International Conference on Performance of Dynamically Loaded Welded Structures. SJ Maddox and M. Prager (eds), July 14-15, San Francisco, Welding Research Council, New York. (1997)
35. Niemi E, Fricke W and Maddox S, Fatigue analysis of welded joints - Designer's guide to the structural hot-spot stress approach, Cambridge, Woodhead, 2006.
36. Fricke, W, IIW Recommendations for the Fatigue Assessment of Welded Structures by Notch Stress Analysis, Woodhead Publishing Ltd., Cambridge, 2012.

GPSR Compliance

The European Union's (EU) General Product Safety Regulation (GPSR) is a set of rules that requires consumer products to be safe and our obligations to ensure this.

If you have any concerns about our products, you can contact us on ProductSafety@springernature.com

In case Publisher is established outside the EU, the EU authorized representative is:

Springer Nature Customer Service Center GmbH
Europaplatz 3
69115 Heidelberg, Germany

Batch number: 10682325

Printed by Printforce, the Netherlands